BABY ANIMALS

BABY GRIZZLY BEARS

by Julie Murray

Cody Koala
An Imprint of Pop!
popbooksonline.com

Hello! My name is Cody Koala

This book is filled with videos, puzzles, games, and more! Scan the QR codes* while you read, or visit the website below to make this book pop.

popbooksonline.com/baby-bears

*Scanning QR codes requires a web-enabled smart device with a QR code reader app and a camera.

abdobooks.com

Published by Pop!, a division of ABDO, PO Box 398166, Minneapolis, Minnesota 55439.

Printed in the United States of America, North Mankato, Minnesota.

102023
012024

THIS BOOK CONTAINS RECYCLED MATERIALS

Cover Photo: Getty Images
Interior Photos: Shutterstock Images; Getty Images
Editors: Elizabeth Andrews and Grace Hansen
Series Designer: Candice Keimig

Library of Congress Control Number: 2023938789

Publisher's Cataloging-in-Publication Data

Title: Baby grizzly bears / by Julie Murray
Description: Minneapolis, Minnesota : Pop!, 2024 | Series: Baby animals | Includes online resources and index.
Identifiers: ISBN 9781098245238 (lib. bdg.) | ISBN 9781098245795 (ebook)
Subjects: LCSH: Animal babies--Juvenile literature. | Animals--Infancy--Juvenile literature. | Bears--Juvenile literature. | Grizzly bear--Juvenile literature. | Grizzly bear--Behavior--Juvenile literature.
Classification: DDC 591.39--dc23

Table of Contents

Chapter 1

Bear Cubs

Baby grizzly bears are called cubs. Females give birth to one to four cubs at a time. Cubs are born in a den in January or February. They stay there with their mother until spring.

Watch a video here!

Cubs weigh about one pound (0.45kg) at birth. They are blind, hairless, and toothless. Cubs are helpless. They rely on their mother to care for them.

Female grizzly bears most often give birth to two cubs at a time.

Cubs sleep, eat, and grow during the winter months. They drink their mother's milk every few hours.

Chapter 2

Out of the Den

Cubs **emerge** from the den in spring. They weigh about 8 pounds (3.6kg). Their hair is brown or tan. They stay close to their mother as they **explore**.

Learn more here!

Cubs are playful. They run, tumble, and wrestle. At night, they go back to the

den for safety. They do this until the weather is warmer.

Mother bears are very protective of their cubs.

Chapter 3

Growing Up

Cubs stay with their mother for two to four years. During this time, their mother teaches them how to survive on their own.

Only about 50% of grizzly bear cubs will survive their first year of life.

Where Grizzly Bears Live

Explore links here!

Cubs learn how to find **shelter**. They also learn how to protect themselves from danger. Mountain lions,

wolves, and adult male grizzly bears are **predators** of bear cubs.

Cubs also learn how to find food and catch fish from their mother. Grizzly bears

enjoy eating grasses, berries, fruits, and roots. They especially like salmon!

Chapter 4

On Their Own

Adult grizzly bears are big and strong. They can weigh more than 900 pounds (408kg)! They are ready to live on their own. Grizzly bears can live up to 25 years in the wild.

Complete an activity here!

Making Connections

Text-to-Self

Imagine you saw a mother grizzly bear and her cubs in the woods. Why shouldn't you get close to the bears?

Text-to-Text

Have you read a book about a different kind of bear? How was that bear similar to or different from a grizzly bear?

Text-to-World

Grizzly bear cubs rely on their mothers for food and safety. What other animals rely on their mothers after birth?

Glossary

emerge – to rise from; to come into view.

explore – to travel across or through in order to discover or search for something.

predator – an animal that hunts other animals for food.

shelter – a place that gives protection against weather or danger.

Index

Online Resources

popbooksonline.com

Thanks for reading this Cody Koala book!

This book is filled with videos, puzzles, games, and more! Scan the QR codes* while you read, or visit the website below to make this book pop.

popbooksonline.com/baby-bears

*Scanning QR codes requires a web-enabled smart device with a QR code reader app and a camera.